Confirming The Safety Coach®

"Dave Sarkus is in your face with *The Safety Coach*®! This book expresses his enthusiasm for safety perfectly, and you will be a believer too by the time you finish. He's infectious and if you don't watch out you'll catch his enthusiasm!"

— Sam Seabright, PE, BSME, MBA, MS
Siemens Automotive Corporation
Manager of Technical Safety • N. America, S. America, and Asia

" *The Safety Coach*® captures simple ideas and introduces them in an entertaining way so managers can immediately use them."

— Dr. Robert J. Kainz
Senior Manager Pollution Prevention & Life Cycle Management
DaimlerChrysler Corporation

"This book will inspire you, teach you, remind you, and help make your leadership a whole lot more effective."

— **Dr. Thomas Stephens**
Professor Emeritus
The Ohio State University

"This book is a *must read* for any leader who wants to easily improve safety through better communications with workers."

— **Wendy Dankovchik**
Senior Manager, Safety, Health
& Environmental Affairs
The Boeing Company
Kennedy Space Center, Florida

"Leaders in any area of business are not simply born but need sound principles like those addressed in this book. David's work will deservedly attract a wide range of readers."

— **Mark A. Perriello, CIH, CSP**
Director Occupational Safety and Health
Environmental Affairs • CBS Corporation

"Great safety performance requires a hands-on approach. *The Safety Coach®* captures the kind of day-to-day leadership that's necessary to bring about world-class safety performance. David Sarkus has created a REAL WINNER!

— Kathleen Harer, PE, CSP
Past President
Board of Certified Safety Professionals

"This book offers leading ideas to improve safety performance from a leading safety advocate, David Sarkus."

— Dr. Earl Blair, CSP
Associate Professor • Safety Management
Indiana University • Bloomington

"Long term change and success comes about by consistently using the fundamental principles like those outlined in this book — excellent!"

— Mark Eliopulos
Division Safety & Health Manager
Kaiser Aluminum & Chemical Corporation

THE SAFETY COACH®

UNLEASH THE 7C'S FOR WORLD-CLASS
SAFETY PERFORMANCE!

COPYRIGHT © 2001
DAVID J. SARKUS

David J. Sarkus

THE SAFETY COACH®

UNLEASH THE 7C'S FOR WORLD-CLASS SAFETY PERFORMANCE!

CHAMPIONSHIP PUBLISHING
DONORA, PA

Copyright © 2001 by David J. Sarkus

All rights reserved under International and Pan-American Copyright Conventions. Absolutely no portion of this book may be reproduced in any form, by any means, whatsoever, without written permission from the publisher.

Library of Congress
Cataloging-in-Publication Data
2001126311
Printed in the United States of America

ISBN 0-9708076-3-5

Fourth Printing 2006

Cover Design: Dennis K. Lange

Illustrations: Cartoon Resources, Inc.

Photography: McLaughlin Studio
Belle Vernon, PA

CONTENTS

Acknowledgements	VII
About the Author	IX
Foreword	XI
Preface	XIII
Introduction	15
Prologue	17
Section I: The Power of Reinforcement	23
Chapter 1: Confirming	31
Chapter 2: Correcting	41
Section III: The Power of Relationships	55
Chapter 3: Caring	61
Chapter 4: Collaborating	69
Chapter 5: Coaching	77
Chapter 6: Conciliating	83
Section III: The Power of Information	97
Chapter 7: Clarifying	103
Bringing it Together	110

ACKNOWLEDGMENTS

First and foremost, I want to thank my friend, *James Malinchak* who inspired me and pushed me to start and finish this book. Thanks James! And I want to thank my parents for their continued inspiration.

I want to thank all of the people who helped me early in my career as a student and later as a professional. At West Virginia University, *Dr. C. Everett Marcum, Dr. Dan Della-Giustina* and *Dr. Jim Weber.* During my early years at COMSAT, *Denver Graham,* who hired me and believed in my talents. At the Kennedy Space Center, *Jay Wortman, Jake Dixon,* and the late *Len Poelker.*

From my days in California, at United Airlines, *Bill Applegate, Rich Petraitis, Joan O'Neil, Scott Robinson, Bill Geheraty, Jim Brenneman,* and *Dr. Bill Wells.* That was a great team at United! At TRW Corporate, *Dave Lavalette* and *Dave Pivarnik* who supported my efforts even those that at first appeared a bit divergent. Also at TRW, *Chris Smith* who hired me and *Tom Pochylski* who remains a loyal friend, confidant, and trusted professional. Thanks for everything, Tom! Also, *Jack Maiuri, Rudy Palone, Bill Gibbs, Nick Nickol, John Kitching, Kevin Witt* and all my other friends in operations and facilities support, at TRW.

In this new phase of my career, I want to thank *Dave Johnson,* the Editor of Industrial Safety & Hygiene News and my friend *Dr. John Kamp* at Reid Systems.

I also want to acknowledge *The Reverend Robert Duch* for his inspiration and writings that led me to the work of *Robert Greenleaf* and to the Greenleaf Center. Servant Leadership remains an integral part of my thoughts and labor as it relates to leadership in business today. In fact, it remains at the core of this book.

For those who initiated my more formal interest in safety and psychology, *Dr. E. Scott Geller, Dr. Tom Krause,* and *Dr. Judy Komaki,* thank you! From the recent past, at St. Mary's College, I can't forget *Dr. James Alan Temple,* a great mentor and dedicated academic who has meant so much to me, helping to refine many of the principles presented in this book. Jim, thanks for everything! I also want to recognize all of the researchers and academics who toil without being acknowledged but should know that they are an inspiration to many.

Regarding my career as a speaker, I want to thank *Bruce Wilkinson* for his unselfish support and for my introduction to the National Speakers Association. Thanks again, Bruce!

Finally, I thank Jesus, the suffering servant, for without him, I am nothing!

ABOUT THE AUTHOR

David Sarkus has been in your shoes. As a health and safety professional, David has had a history of success helping others improve safety performance in every position he's held in his 20-year career. He knows his way around and has been "in the pit." David knows how to get his hands dirty. Most of his work has been in large Fortune 500 settings with corporations such as United Airlines and TRW, Inc., and at NASA's Kennedy Space Center. But he has also worked with smaller organizations.

David is the president and founder of David Sarkus International, a training and consulting company that specializes in strategy, leadership, and behavioral safety performance. He is a nationally recognized author, speaker, and consultant. David holds a master of science degree in organizational psychology from St. Mary's College of California and a master of science degree in safety management from West Virginia University. He's also the technical editor of *Industrial Safety & Hygiene News,* an international safety publication.

David Sarkus is recognized as an outstanding motivational speaker who regularly helps companies keep their people excited about safety! He speaks at individual-site safety gatherings, group and regional events, and corporate conferences.

BOOK 'EM

If you want to bring the *7C's* and other themes alive in your organization, David can deliver his message with an engaging blend of experience, humor, storytelling, and passion. He can be reached by calling or writing:

724-379-6439 • 800-240-4601

Other contact information:

David Sarkus International

www.DavidSarkus.com

THIS BOOK

Your organization can purchase this book with special volume pricing when you bring David in to speak to your group.

FOREWORD
The Power of The 7C's

As a young trainmaster, many years ago — the very first week I ever worked as a manager, a Saturday came along, and we had a fatality. I was the only manager remaining in town — all of the senior managers had left. I was absolutely petrified about what to do and how to go about processing this horrible accident, appropriately. Since then, I've had numerous situations where I've been involved in tragedies, and have learned that we can prevent all of these injuries, and all of these accidents.

For decades, we've had programs that varied from one year to the next, often becoming the "flavor of the month." So during my career, I've tried to develop and implement safety processes that would give us a bit more traction. With this in mind, Don Seil, one of my colleagues, was looking for some different ways to more deeply integrate safety into California Northern Railroad. Don ran across this book, I read it, and called him back, and said, "You know Don, there's only one thing wrong with this book," he replied, "What's that?" and I said, *"I should have written it!"*

At RailAmerica, we've worked together to push our organization up to "Championship Level." And the bottom line, the principles that I believe in, especially the engagement process, and working with people, and caring, and all of the other principles that David Sarkus writes about in this book are things that *I know work!*

We've adopted "The 7C's" and I endorse them as a very effective process. These seven principles will work for your organization as well — and they will help boost productivity, too!

Respectfully,

Bob Jones, West Region Vice President
Chairman Cross-Regional Safety Team
RailAmerica

More About Bob Jones and RailAmerica: Bob Jones attained the American Short Line and Regional Railroad Association's (ASLRRA) highest individual safety honor as the *"2008 — Safety Person of the Year."* Jones led RailAmerica to its safest year ever in 2008, with a 1.64 FRA frequency index, and a 35%, year-over-year improvement. In 2007, Jones customized a safety program based on David Sarkus' book "The Safety Coach®: Unleash the 7Cs for World-Class Safety Performance." and "The Safety Coach® Learning System™." Jones embedded the 7C's into the region's safety culture and helped implement the concepts at all RailAmerica properties.

PREFACE

This book is based on the premise that if safety-related behaviors and underlying attitudes can be improved, the number of accidents will be greatly reduced. And, if we can all become better SAFETY COACHES — influencing people in positive, non-manipulative ways, really caring for their well-being, great outcomes will follow.

At some level, safety is everyone's responsibility. But at the core, management is largely responsible for creating the right climate where safety will be supported in ways that are comparable to quality and productivity. By using sound principles like those presented in this book, the challenge of creating an injury-free culture will no longer be viewed as an impossibility but rather embraced as a very real possibility.

This book is for everyone from the line manager, plant manager, director, vice president or general manger, the mechanic, machinist, you name it; it can help every person to become a better leader and COACH.

Whether you need to improve a given program or process, or start something new, this book will help to support without displacing what you've already created. And it can be used most effectively when people are held accountable to use these principles on a regular basis.

INTRODUCTION

At times, it's easy to think that safety isn't that important and should be left for the "safety people" to do. But you also know that everyone has a role to play when it comes to improving safety performance. And I know that you want to play your part and make good things happen!

There's little excuse for *not* making this book work for you. THE SAFETY COACH® is an easy read and can help you improve your culture for safety — right away. The seven principles that make up this book provide a valuable antidote to some of your daily safety challenges. Just take a little time to use them, every day.

Inside you'll find solid content that will help you to positively influence people toward working more safely and more productively. The principles were developed from my safety-related efforts within a variety of industries and throughout my graduate studies. The principles are deep and complex, yet they've been made very practical.

This book can show you and others how to become a great SAFETY COACH®, and to help everyone avoid costly and painful injuries. In a very concrete way, I've used over 20 years of my experience to help people like you — help others work more and more safely, both on and off the job.

The Safety Coach®

Many safety-related books do not provide an easy way to implement *whatever* is being taught. You won't have that problem with this book — it's clear and concise. The content can easily become a part of a given program or process that you may want to initiate or improve. It's a book that provides powerful principles that will make you a winner! THE SAFETY COACH® gives you great principles — powerful stuff that can help you achieve World-Class Safety Performance!

These positive and powerful principles are contained in the *7 C's:*
Confirming, Correcting, Caring, Coaching, Collaborating, Conciliating, and Clarifying.

DON'T WASTE TIME - GO OUT THERE AND CATCH' EM WORKING SAFELY!

PROLOGUE

To help introduce the *7C's* and explain the three main sections of this book, I want to tell you a true story about Willie and Joe (their names have been changed to protect the innocent). Willie was somewhat of a problem employee who was transferred to a department that was effectively using a behavior-based approach to safety. His mentor was Joe, a co-worker and safety observer well-liked and respected for his skills as a machinist, and as well as his strong convictions about safety.

Willie worked closely with Joe, day in and day out. Over the course of six months to a year, Willie's safety-related attitudes and actions changed. At first, these changes were pretty superficial. He acted or thought in certain ways only to gain favor from Joe.

Later, he worked safely because he knew that there were mutual expectations within his new group, as well as with Joe. He really did want to maintain good relationships with his new colleagues.

In the end, through constant and persuasive feedback, Willie realized that working safely enabled him to enjoy the things he valued in life.

What's going on when you see people like Willie experience these kinds of positive changes? And more practically speaking, how do you bring about broad changes in attitudes and actions — changes that go beyond a narrow set of targeted behaviors? How do you sustain these new attitudes and behaviors without constant forms of small rewards, praise, or various forms of peer pressure?

Having worked with Willie and Joe, I want to describe three dynamic processes that occurred in their story. The first focuses on a personal form of superficial compliance that comes from the CONFIRMING and CORRECTING feedback of a coach.

THE POWER OF REINFORCEMENT

Let's go back to Willie's first few months in this new department. Initially, his attitudes and behaviors were mainly associated with social effect. Willie wanted to be rewarded by Joe, and to avoid being punished by his supervisor for non-compliant behaviors.

Publicly, he looked good, but when nobody was around Willie usually reverted back to his old "unsafe self." Joe proved very wise here. He CONFIRMED Willie when he was working safely through small rewards and praise, but didn't CORRECT him nearly as much. You see, the use of small rewards (at times a token for a soft drink or some food) and praise increased Joe's personal appeal. This set the stage for a somewhat deeper and more permanent form of influence through "The Power of Relationships."

Prologue

THE POWER OF RELATIONSHIPS

After several months, Willie was changing a bit more, behaving safely even when others weren't around. He wanted to maintain the relationships he was building within this new group, and in particular, his relationship with Joe. There were mutual obligations that had to be met, and he was smart enough to realize it. For Willie, the mere "psychological presence" of Joe and the others he respected was enough to keep him working safely. Not all of the time, but much more often than before.

This second process draws on CARING, COACHING, COLLABORATING, and CONCILIATING. Joe was seen as a credible and trustworthy co-worker who wanted to help Willie. He became a trusted friend because he CARED for Willie by doing work-related favors and by listening; he COACHED Willie by setting a good example; he COLLABORATED with Willie by making him an active participant in the safety process; and he resolved conflicts before bad feelings could take root.

Somewhat unknowingly, Joe was motivating Willie to maintain this ongoing relationship, and to meet the expectations that went with it. If Willie wanted to keep this work relationship positive, he knew that working safely was one way to do it.

Emotional ties were being formed. Willie and Joe were building a strong and positive rapport, largely because Joe was a well-respected co-worker and trusted friend. All of this helped lead to an even deeper phase of change through "The Power of Information."

THE POWER OF INFORMATION

The power of information was the third process at work in Willie. Over the course of a year, Willie began to realize that safety was personally important, and he became fairly open in sharing these thoughts and feelings. This awareness and expression is the thinking phase of the process, and it's helped along by someone who CLARIFIES these perceptions and values.

At first, Willie spoke about how he liked being rewarded by Joe. He recalled how he also wanted to maintain a good relationship with Joe. But overall, it was Joe's positive and persuasive feedback that won him over. Joe often told Willie that sooner or later he would get hurt if he didn't change his habits. Willie could avoid injury, he explained, by realizing that he was in control and could take appropriate actions.

More importantly, Willie experienced inconsistencies between his values and his actions. He began to acknowledge that working safely meant he could continue to engage in his favorite activities outside of work with family and friends.

Through all of this, Joe helped to CLARIFY the individual relevance of safety for Willie. Willie became a leader in his own right, no longer needing needing to be with or around Joe in order to work safely. The changes in his attitudes and behaviors were deep and lasting.

Even after years of downsizing and "rightsizing," mid-level and senior managers at Willie and Joe's plant continue to realize the bottom-line benefits of nurturing these three processes — based on reinforcement, relationships, and finally information. They support this interactive method as an ingrained part of the culture. And the workforce remains committed to influencing others like Willie, because they all have experienced the benefits of the transformation themselves.

"Feedback Is The Breakfast Of Champions."

— Rick Tate

SECTION I
THE POWER OF REINFORCEMENT

SECTION I
THE POWER OF REINFORCMENT

Positive communication can be neglected when it comes to safety. For some reason, our history leads us to believe that we have to be heavy handed and practice tough-minded principles at every turn.

Remember, for individuals to work at their most productive levels, or in the safest ways possible, they need regular feedback about how they're doing. People have to receive some type of verbal or visual information regarding their performance. This is especially true when it comes to safety. If workers don't get regular feedback that's helpful or positive, they'll simply work in ways to avoid some kind of punishment or negative response.

But it's human nature to work better when feedback is positive. For this reason, managers, supervisors and co-workers benefit most by using positive principles that will help people increase their desire to work safely, instead of just trying to avoid punishment from others. When individuals and groups receive helpful information about working safely, they respond by working more and more safely. It's that simple!

WE ALL NEED FEEDBACK

Be Yourself and Be Positive!

OVERT COMPLIANCE

Think about this scenario; you've likely been there before. You're wearing your personal protective equipment (PPE), maybe a hardhat and eye protection. You're also in some kind of supervisory role and people usually follow your lead. You walk over to a group of people who were working alone for some time, and now, at least some of the workers begin to look for their hardhats or glasses and start to put their PPE on. After you leave, you notice that they begin to take off some of their equipment, knowing that they did what they thought they had to do while you were there! The actions of those workers weren't very positive or deep. Actually, they were pretty superficial, weren't they?

If you don't actively help others to work safely, people will tend to work at a superficial level of compliance. Superficial compliance simply means that workers will do what's expected of them when others are around, if they're being pushed to work safely, or if they're being watched. But when the boss or some other person of importance leaves the area, they often revert back to some "less-than-safe" type of behavior. In a way, they're trying to gain a type of reward or to avoid some kind of punishment from their co-workers or supervisors!

UNDER SURVEILLANCE

Does This Seem Familiar?

TAKE A MINUTE

Does the following industry practice regarding safety apply to your company? Workers only hear from their manager or supervisor when someone's injured or if something's damaged. Even worse, most people work any way that they can to get the job done, knowing that their supervisor will usually look the other way or never notice how they're working.

It's critical to talk to your people about safety every day. One minute spent with a person talking about safety can be a life-changing event. I'm sure you know of someone who, after hearing about or seeing an accident or bad injury, talked about how he wished he would have done something or said something to the injured worker. The person may have mentioned "I knew something was going to happen" or "I wish I would have said something to him earlier, he didn't have to get hurt!"

One minute spent talking about working safely could be worth more than all the money in the world! Think about it the next time you fail to discuss the importance of working safely with the people around you.

SOCIAL PUNISHMENT

The Boss Has Not Been Around In Weeks!

> **"The sweetest of all sounds is praise."**
>
> — Xenophon

CHAPTER 1

CONFIRMING

CONFIRMING AS REWARD

Remember, rewards come in all sizes and shapes and don't necessarily have to be a gift or money. It can be a smile, a thank you, a handshake or even a nod of approval, and the sooner the reward is given *after* the behavior, the better. These are social rewards. Catch your people doing something safe and then CONFIRM that behavior! CONFIRMING a safe action means saying, "YES" to what they're doing safely! It means that you're applauding their performance and want to see more of it on a regular basis.

When it comes to positive feedback, CONFIRMING, make use of the 3 F's:

- Fast, CONFIRM right *after* you observe a safe action;

- Frequent, do it as much as possible; and

- Favorable, make it as positive as you can.

People are better able to accept feedback that they know has appropriate meaning. That's why it's important to be yourself and to be genuine!

EVERYDAY EVENTS

Be a Hero-Maker!

CONFIRMING IS GOOD

BY "CONFIRMING in favor of CORRECTING," a number of real good things can happen. You can begin to move people away from mere compliance towards deeper levels of personal commitment. For one, you can help your people find satisfaction in working safely. By reinforcing the safe actions of the people around you, individuals can begin to see that working safely is important not only to you, but also for them.

By being helped to work safely and stay whole, workers can gain a healthy respect for their supervisor, boss or lead. They can begin to realize that others are there to help rather than hinder their work.

With this basic principle, CONFIRMING, you can become more valuable to the organization, not just in terms of safety, but also with regard to morale, productivity and quality. You can help to build respect and trust throughout your organization. Finally, you can increase your own positive power by becoming more attractive, psychologically speaking!

BECOMING ATTRACTIVE

You Can Become More Attractive Too!

DO IT DAILY

There are many actions that can be CONFIRMED on a daily basis. Think about a few. The actions might include the use of certain kinds of PPE such as eyewear, fall protection, hearing or hand protection. You can also praise workers for following certain procedures, such as the safe operation of a forklift or another piece of machinery. The possibilities are endless.

Look around — there are many more opportunities to start CONFIRMING in favor of CORRECTING. Take a look at some of your work operations:

- Hazardous materials usage
- Material handling and storage
- Wood or metalworking operations
- Use of hand and power tools
- Fire prevention practices

Take a closer look at the activities of those around you so you can help to make them safer. CONFIRMING can become contagious and transform your organization!

TAKE A MINUTE

With 60 Seconds In It!

BE PERSISTENT

People want to work safely and, when others CONFIRM these actions they begin to feel good about a job done well *and* done safely! By consistently CONFIRMING the safe actions and work of your people, you can move your organization to levels of performance you never imagined! Levels that will gain attention and notice. Levels that can move your firm toward zero accidents!

Remember, be consistent. Don't just CONFIRM the safe actions of others when you feel like it. You can't allow your feelings to dictate whether or not you will help to reinforce the safe actions of those you work with and around.

If you can learn to CONFIRM in favor of CORRECTING, safe actions will almost always follow. And when safe behaviors begin to increase you're reinforcing the kinds of actions that can keep people from becoming ill or getting hurt. But it takes persistence and dedication to deliberately go out and work at CONFIRMING safe actions, regularly.

CELEBRATE SUCCESS!

People Can Get Excited About Their Achievements!

> "It is an art to drive hard with a light hand."
>
> — Wise Old Woman

CHAPTER 2

CORRECTING

REDIRECTING OTHERS

Sometimes, workers may not know the proper or expected procedures within their organization. They might not have the right tools or equipment to perform the job as required. Perhaps they don't have the right protective equipment to do the job safely or it may be difficult to obtain. They may lack the physical or mental skills to do the job safely. Any one of these reasons can present an obstacle to working safely. If any of them are true in your organization, they'll need your help.

Start by asking questions of that person or their co-workers. If there are obstacles like those described above, work as hard as possible to remove them. If there are no obstacles and the person simply doesn't want to work safely, then you've got a bit more work to do.

CORRECTING does have its place in safety and can serve as a form of punishment by decreasing unsafe or at-risk work activities. Feedback that's CORRECTING should be used when people are consistently found to be working in some at-risk way. But first, it's often helpful to explore why someone is working in an at-risk manner.

APPROPRIATE SUPPORT

Your People May Need a Little Help!

HELPFUL CORRECTION

Most of the time, CORRECTING feedback can be handled in a very constructive way. CORRECTING and redirecting almost any type of at-risk action in a positive way is a good method to greatly reduce unsafe actions. You can be firm and still be positive.

Let your people or co-workers know that you're trying to help them stay healthy and whole! If it helps, ask for their permission to give the kind of feedback that's needed. You may want to begin by asking a few simple questions or making a couple of statements for example:

- Ask if there might be a safer way to perform the job

- Ask why someone may not be doing the job as safely as it can be done

- Explain that others have been injured by working in the same way.

Ask questions and leave room for a two-way exchange of information and you can open the door for positive changes in safety performance.

CREATING CONSEQUENCES

Be Clear and Concrete!

CORRECTING FEEDBACK

As with CONFIRMING feedback, be specific about the kinds of behavior that are placing people at-risk for injury. Don't simply tell people you want them to "work safer." For example, you can point out *specific* actions they could take relative to fall protection, eye protection, or driving safety. Give constructively CORRECTING feedback as soon as possible, especially when a serious unsafe action has been observed.

Be cautious about using this principle; don't overdo the CORRECTING! As you read before, there are many more opportunities to CONFIRM rather than CORRECT. If you only CORRECT, you won't increase the kinds of *safe* behavior you want to see. You may do away with some unsafe actions for a period of time but you need to replace those actions with safe behaviors that can be reinforced — CONFIRMED!

Some people feel they shouldn't have to praise workers for doing something that is expected or required. But working safely, day in and day out is difficult, and people need support from individuals like you to keep them moving in the right direction.

BE SPECIFIC

Be Specific About What Is Expected!

BE PATIENT

Without a doubt, it can be hard to work safely every day. It can take more time, effort, and may be a little uncomfortable. Be patient with your people. CORRECTING, more than CONFIRMING, strips people down and leaves them feeling naked, without pride and purpose. Constant CORRECTING can become the equivalent of human correction fluid, wiping people out! Don't over-CORRECT! Also, you become less attractive psychologically to your workers or co-workers.

CORRECTING feedback can be handled in a very constructive way. CORRECTING and redirecting some type of unsafe or at-risk action in a positive manner is a good method to stop unsafe behaviors. By CORRECTING in a helpful way, you can remove the unwanted side-effects that can result when feedback is given in a harsh, overly critical, or angry sort of way. You can be firm and still be positive! Let your people know you're trying to help them stay safe and whole. Don't forget, once safe actions do begin, go back and CONFIRM those behaviors in order to "fill in" the empty spot.

DON'T OVER-CORRECT

Over-Correcting Can Create A Void!

DON'T PANIC

Sometimes workers may just be trying to push your buttons by purposely doing things they know will annoy you. You will still need to correct and redirect but you can also pay less attention to them. If you've figured out that some employees are trying to get you bothered, just ignore them for a good long while. Cool them off by ignoring them. It works — but don't panic, they may try even harder to upset you if they fail to get the desired reaction.

Remember, peer pressure is a powerful thing, and in time, problem employees will begin to match the safe actions of the group. When people are ignored, they have to begin to figure out what to do next. Is it going to be worth their while to keep playing games? Probably not if they are reasonable people. Will they determine that it just might make sense to go along with positive change, working safely and being supported by those around them? Usually this is the case.

Hang in there. By ignoring some of the more challenging employees, you can extinguish undesirable actions and begin to help them see the kinds of behaviors that are best for everyone.

JUST IGNORE THEM

Learn To Extinguish Undesirable Actions!

DAVID'S SAFETY TIPS

- Remember, consistently CONFIRM in favor of CORRECTING.

- Be specific when talking about the actions you're CONFIRMING or CORRECTING. As much as possible, give your feedback right *after* the action is observed.

- CORRECT and re-direct constructively but don't OVER-CORRECT.

- Most types of CORRECTION should be followed-up with CONFIRMING feedback.

- Difficult employees may have to be ignored for a good long while until they conform to the kinds of positive changes that are benefitting the group.

YOUR NOTES PAGE

> "Effective Change is not something you do to people.
>
> It's something you do <u>with</u> them."
>
> — Ken Blanchard

SECTION II
THE POWER OF RELATIONSHIPS

SECTION II
THE POWER OF RELATIONSHIPS

No matter what people may lead you to believe, almost everyone wants to maintain good relationships at work. Through the power of relationships, people can begin to work safely in a variety of ways you never expected. They'll begin to work safely even when others aren't around! Usually, this is done in order to maintain or improve important relationships.

When workers want to uphold good relationships with their boss, supervisor or co-worker, just the thought of that person is enough to keep them working safely. The thought or "psychological presence" of another person can move people to take the right kind of actions — safe actions!

Relationships serve as a powerful force in helping others to work safely. As many people can attest, relationships are as important as any program or process and must be cared for and nurtured.

PSYCHOLOGICAL PRESENCE

POSITIVE RELATIONSHIPS

By maintaining and improving positive relationships, you can help people to work safer in a number of different ways. To do this, you'll need to start using the next four C's: CARING, COLLABORATING, COACHING and CONCILIATING. By using these C's, safety performance can begin to move even further away from compliance and closer to real commitment.

It's been said that the true measure of our influence is how people act when we're not around to watch them. And much more of our influence — not only as supervisors or managers, but as peers — is based on the places we occupy in the hearts and minds of those who work for and with us.

CARING leaders are trusted. They are consistent in both word and action. This creates good working relationships that can be embraced and even cherished.

Good relationships allow us to lead more effectively on the basis on commitment, not just mere compliance. The best form of positive influence comes from conveying a sense of CARING; call it "CARING-leadership."

THE MERE THOUGHT

The Power of Caring-Leadership!

"Caring for persons, the more and less able serving each other, is the rock upon which a good society is built."

— Robert K. Greenleaf

CHAPTER 3

CARING

RELEARNING CPR

Any person who helps others to work in safer ways is a leader. However, to be truly effective you need to learn and practice "CARING-leadership." Now to do this you have to relearn **CPR.**

You're saying "CPR, cardio-pulmonary resuscitation?" I say, "no, no, no!" I'm talking about **C**ARING-leadership, demonstrating **P**rofessionalism in all you do, and showing **R**espect towards each and every individual.

By practicing this form of CPR, you can breathe new life into your organization and people can get excited about improving safety!

Healthy organizations show CARE and concern by providing valuable tools and learning experiences that help employees openly and positively influence the work habits of others. By having a healthy concern shown for them, people learn what behaviors are expected and how to provide feedback. When communication and influence are founded on a genuine desire to help, such efforts aren't viewed as coercive or manipulative. When people CARE for each other through safety, a sense of community can be established in an organization and its members can feel a *spirit* of support!

CARING-LEADERSHIP

A New Form of CPR

EMOTIONAL TIES

CARING for the people around you can develop strong emotional ties within your organization. These ties can naturally evolve into tendencies for workers to want to help each other and to do their best. CARING leaders know how to build and maintain solid and substantial relationships:

- They allow for two-way feedback and really want to know what others think.

- They share the blame for failures that may have occurred and don't point fingers.

- They keep their promises or give valid reasons when commitments can't be kept.

- They recognize the needs of individuals and know that everyone has a valuable part to play in the organization.

Do you want to demonstrate CARING-leadership? You've got to be willing to get to know the people you work with. Be open to understanding them as individuals. Get to know what they like and don't like. Get to know them by name, as a person, and not just as another employee.

CARING IS PERSONAL

Get To Know Those Around You!

CARING FOR OTHERS

CARING-leadership should serve as the foundation for all we do. It begins with a heartfelt sense and desire to help others to do and be their best. If we don't truly CARE for the well-being of those around us we're missing the boat! But if we do CARE, our actions will show our concern.

Take a few minutes and think about the ways you can show CARE and concern for your people or co-workers. The possibilities are endless.

You can attend your group's safety meetings and greet people with a smile and handshake.

You can lend a helping hand with a difficult task.

You can make a hospital visit or visit the home of an ailing employee.

You can be available to listen to the challenges that your workers face at home or work. I'm talking about really listening. Listening with your heart!

In many ways, CARING is the starting point for helping others to work safely. We learn to model ourselves after people who go out of their way to help and CARE for those around them.

YOUR BASE

Caring – A Foundation For Leadership!

> "Active involvement can become a part of your winning formula for success — Collaboration!"
>
> — David Sarkus

CHAPTER 4

COLLABORATING

ACTIVE INVOLVEMENT

The second way for leaders to improve relationships and increase the tendency of others to work more safely is by COLLABORATING. In short, COLLABORATING means involving workers in meaningful ways to improve safety.

Many organizations know that when people are regularly engaged to help the organization succeed, there's a greater sense of pride and ownership, especially when good things are accomplished. Workers need to be given the knowledge and trust to succeed or even fail. Through COLLABORATION, people will want to reach new levels of achievement in safety!

Active involvement allows workers of all types to come up with ideas and solutions that are more powerful than those from any one person. COLLABORATING among all types of employees, managers, directors, and hourly personnel can bring people together in new ways. Then they can move forward in concert with less resistance. This kind of togetherness may take longer to foster — but ultimately it leads to an increasing number of positive changes.

PEAK PERFORMANCE

Collaboration – A Winning Formula!

ONGOING IMPROVEMENT

Higher levels of achievement will come to be expected, and people will continually want to be involved with improving safety when individuals work together as one group or unit. However, think about what occurs in some organizations. Many times, workers are simply told and directed what to do. They're not listened to, or they're left completely out of an important safety-related process that they know more about than anyone else. When this happens, people begin to believe their work is *not* important. People think or realize they have no power, so they give up. They're left purposeless and powerless!

Most organizations operating this way just can't survive in today's very competitive environment. Today's competition requires that people work together so that performance is high as is trust. High trust organizations expect the best of their people and in return they get the best from individuals and groups. These organizations know that people are an incredibly valuable resource that must be properly nurtured in order to meet the demands of today's global economy.

Chapter 4: Collaborating / 73

PURPOSELESS PEOPLE

Actively Un-Involved!

VISION AND PURPOSE

By COLLABORATING with workers on a regular basis you can empower them in ways that they can get excited about working *together* to improve safety! People can begin to see the importance of their own purpose and mission within the group. They can start to see and move toward a clearer vision for excellence in safety.

Through actively involving others — COLLABORATING — individuals are treated as your company's most valuable resource. And, as with all of the other positive principles outlined in this book, there are plenty of very rewarding ways to COLLABORATE with the people around you.

People can be involved with the selection of new equipment or on a design team for a new process.

You can involve workers in training that will help them to better coach each other to work more and more safely.

People can be involved in designing a vision for safety.

The opportunities are limited only by your imagination!

AN ENGAGED GROUP

"Coaching, more than anything else, is about modeling the way, setting a great example!"

— David Sarkus

CHAPTER 5

COACHING

LEAD THE WAY

The third way to maintain and improve positive relationships and to help others begin working safely in a number of varied ways is through COACHING. As much as anything else, COACHING involves modeling the way, setting a great example for safety! When you model the way through your own safety-related behaviors, your actions speak louder than any words.

If you haven't noticed, COACHING in sports has changed in recent years. To some extent, the dictatorial head coach is vanishing, the one who operates with top-down authority —"my way or the doorway." In business, today's coach operates on more of a horizontal plane. Rather than using top-down authority, he or she is more a part of the group and allows for two-way communication — moving from "my way" to "our way."

Take a moment to think about how you communicate with others. Do you model the way at work by setting a good example when it comes to safety? Are you willing to set higher standards of achievement by holding yourself accountable and doing what you'd like others to do?

OUT IN FRONT

Leading By Example!

ACCOUNTABILITY

The standards you establish by holding yourself accountable to set a great example are critical for improving safety performance. You may need to wear certain kinds of PPE such as eye protection or hearing protection or you may need to show more excitement about safety! You may have to set the tone by correcting hazards quickly, making engineering improvements or purchasing safety-related equipment so people will be well protected. Your actions do speak louder than words and can leave an imprint that will not be easily erased!

High standards of performance are set by great coaches who use their knowledge and experience to help others improve their game — in our case; the game is improving safety performance. The best safety coaches have a strong desire to bring out the best in others so workers can believe in themselves, and know that working safely is in their best interest, and the team's best interests. This is all a part of higher standards and expectations.

Great coaches need to be consistent yet flexible in their dealings with others, allowing potential leaders to develop from within the group, pushing performance up another notch.

A NEW LEVEL

Keep Raising The Bar!

> "If you hold a grudge dig two graves."
>
> — Ancient Proverb

CHAPTER 6

CONCILIATING

MAKE THINGS RIGHT

The fourth principle that will help to maintain and improve positive relationships, while helping others to continue working safely, is CONCILIATING. CONCILIATE means to resolve disputes or conflicts so that relationships can be repaired. So that people will continue to communicate, CONFIRMING in favor of CORRECTING and use the rest of these positive principles. If disputes are not resolved, communication will decrease or even stop, and safety performance will suffer. This downward spiral can lead to frustration, fear and lost productivity.

Helping others to resolve unnecessary conflict is the right thing to do and will help to create the right climate for ongoing improvements in safety performance. It takes a big person to admit his or her wrongs, but by doing so organizations can improve by developing closer and stronger relationships that will have a positive influence throughout the organization.

IT CAN BE DIFFICULT

It's The Right Thing To Do!

FIRST STEPS

Let's take a look at several steps that can help you resolve non-productive conflict that hurts all kinds of performance, including safety.

1. Focus on the issue. Before you speak with another person or group about an existing conflict, keep in mind that your perspective may not be as objective as you would like to believe. Give yourself time to let go of disruptive emotions so that you can approach the other person with respect. This allows you to maintain your dignity while preserving the other person's as well.

2. Consider temporarily giving up your position of thinking. This should be done at least for the period of time it will take to reach an initial agreement. Don't contemplate alternatives. As difficult as it may seem, put aside personal differences and focus on the issue and the common cause shared within the organization. With safety, this should be easy to do.

LET GO

Let Off Some Steam!

A BIT MORE

Here are steps three and four for resolving conflict in your organization. Both are designed to create a level playing field for both sides involved in the dispute.

3. Meet on neutral ground. Sometimes it's helpful to choose a location that's not considered your 'turf' or theirs, particularly if a lot of tension exists. Go to a break area, a conference room, someone else's office, or even out to lunch. By moving to another place, it can be easier to let go of bad feelings, tension, or other distractions and interruptions. Also, consider using a facilitator if the conflict is too large and could benefit from a third party's assistance. Finally, remember to leave enough time to get through the process without feeling rushed.

4. Listen up. It's important to hear others first. Listening is vital to resolving conflict — it's "the great equalizer." Emotionally, many people just need to vent before they're able to accept the viewpoint of another. By listening, you demonstrate that what someone has to say is important, important enough to "really listen." While listening, don't allow your mind to wander to prepare your next thought or position. Concentrate on what is being said.

Chapter 6: Conciliating / 89

THE GREAT EQUALIZER

Work on Your Listening!

JUST THE FACTS

Here's step five to help you move closer to resolving conflicts. Your integrity involves sticking to the facts and not delving into personal issues that are not central to resolving the conflict.

5. Stick to the facts. It's often better to describe what actions have been observed rather than telling someone about their attitude. Describe behaviors that have occurred, not personality issues or characteristics. Be specific about what you've observed.

Here's an example: You watch Susan, a maintenance associate, abruptly walk out of a steering committee meeting. You know that something is wrong. Later, you approach Susan and ask her if something was bothering her at the meeting.

Let her know that you saw her walk out of the meeting and that you were a bit concerned. Let Susan know that this is only your opinion by saying, "I saw you leave and I'm a bit concerned..." Emphasize that this is only your opinion.

This is a non-threatening way to open up an exchange and to begin dissolving the conflict so that solutions can be discussed and implemented.

NOT ATTITUDES

Talk About Actions!

CLOSE THE LOOP

Finally, here are steps six and seven for CONCILIATING. These are important parts of closing the loop and ensuring that the conflict becomes more completely resolved.

6. Work for a win-win. Joe, a maintenance worker, was disturbed about the absence of fall protection, particularly about a couple of units on back-order. He didn't know that the manufacturer had trouble completing the order and felt that his own organization was dragging its feet. He didn't think his needs or those of his co-workers were being met. Once his concerns were laid out on the table, the issue became clear and alternatives were addressed. In fact, guardrails were added at the work site and fall protection from another organization was borrowed until the new equipment arrived.

7. Follow through and follow up. Conflict and tension may still be present if the solutions are not further addressed. From time to time, check back and ask questions that permit more communication about the solution. This is an important part of resolving conflict. If the conflict is large, take time to celebrate in some small way and let go of the past.

Chapter 6: Conciliating / 93

BE SURE TO ...

Make Sure You Follow Through!

DAVID'S SAFETY TIPS

- CARING for your people sets the stage for other more positive forms of influence.

- Using each of these four C's: CARING, COACHING, COLLABORATING and CONCILIATING builds trust and strong relationships.

- When consistency and trust are established, your message can be more quickly accepted, leading to positive changes in actions and attitudes.

- Not all conflict is bad; you need to figure out the bad from the good. On the positive side, conflict can help define responsibilities and move people to action. Work toward resolving the bad, let go of the past and move on as quickly as possible!

YOUR NOTES PAGE

> "Lasting change connects the head and the heart – it's an inside job."
>
> — David Sarkus

SECTION III
THE POWER OF INFORMATION

SECTION III
THE POWER OF INFORMATION

To help others work more and more safely, you have to know what actions are needed to keep people injury-free. In other words, you need to know what you're talking about! You don't have to be an expert, but you do have to know the difference between "safe" and "unsafe." By using each of the C's outlined in previous chapters, you will become a person who can be trusted, a supporter of safety. You will be viewed as a powerful source of positive influence!

When you're seen as somebody who really wants others to remain healthy and whole, your message will get through to those who work around you! However, once is not enough. You need to be consistent and persistent! People need ongoing feedback of different sorts to help them work safely.

Don't give up or give in. Keep providing positive reasons for others to work safely. And remember to set a great example by practicing what you preach so that your level of trust will go up and people will listen to what you have to say.

IT IS POWER

Bring Your Knowledge to Life!

THE POWER OF INFORMATION

When you're trusted, your message will be viewed as one that really makes sense, one that will guide workers to make lasting and durable changes in their safety-related attitudes and actions! The light will go on for many individuals as they align their beliefs and values with their actions. Values are those things that we feel are important in our lives; for example, almost everyone values their health and family. Workers can't change their values very easily, but they *can* change their actions, especially if they're hearing the message about working safely on a regular basis. This is when the head and heart get connected!

When you begin to help others CLARIFY the personal importance of working safely, you're getting to the heart of what they really want — what's really important to them!

People need an outside perspective to help them understand more clearly what's important to them and to know that the organization supports their safety. Every individual needs to know that the organization values safety and wants him or her to take the time to value it on a personal level. When the organization values safety, it will become evident through the actions of every individual.

Section III: The Power of Information / 101

A BIG CONNECTION

Moving Toward Lasting Changes!

"You can't threaten or force people to change — ultimately workers convince themselves that working safely is the right thing to do."

— David Sarkus

CHAPTER 7

CLARIFYING

YOU HAVE A PART

You have a big part to play! It's up to you to deliver a consistent message that helps workers understand that they have the most to lose *and* gain when it comes to safety.

What you say can help to CLARIFY what's personally important to each and every worker. By CLARIFYING what's important to each person, you can help them to make changes in those behaviors that don't line up with their values. In some ways, you can become the messenger of hope! Someone that delivers a message that can be accepted, not rejected! A message that promotes consistency between actions and values!

CLARIFYING values relative to safety is not a "cure" or a vaccination against injuries or losses. But it does offer people insight and a way to look at choices that can become a very strong part of who they are and what they stand for.

If you think about it, there really is no place to hide from your values. And when values are regularly CLARIFIED, those nagging and bothersome thoughts to work safely won't go away too easily. People will eventually have to choose those things that they care for most, weaving them into the fabric of their daily lives — on and off the job!

ALIGNMENT

On Track – Values and Actions!

MAKE IT PERSONAL

People need to understand that it's personally important for them to work safely every day! By having brief, one-on-one COACHING sessions, you can help people understand how important it *really* is to work safely, especially once you get to know them as a person. In group meetings you can use graphs to chart safety progress and CLARIFY the specific kinds of action that will bring them closer to their goals. You can also ask workers to identify the most important things in their lives and prioritize that list. Workers can be given time to provide their personal testimony regarding an accident or close call with a discussion of how to avoid similar situations in the future.

Beyond what has already been written, part of the purpose in CLARIFYING values is to help people to: 1) Find purpose in working safely. To know that working safely is not a waste of time but is very meaningful; 2) Become more productive so that working safely can give them a sense of accomplishment in doing a job well; 3) Sharpen their critical thinking when it comes to safety and at times explore safe alternatives if challenges exist; 4) Improve relations with others who have common goals and interests — valuing safety on a personal level!

YOU ARE INDEED THE AGENT OF POSITIVE CHANGE — MAKE IT HAPPEN!

STORIES HAVE POWER

DAVID'S SAFETY TIPS

- Working safely needs to make personal sense for every worker. They need to know that they make decisions and take actions that are very important to their well-being.

- A message delivered from a trusted co-worker, supervisor, manager or family member can be critically valuable in moving others toward a strong and durable commitment to safety.

- You can CLARIFY the personal importance to work safely through group and one-on-one discussions that will help individuals better understand their values as well as whether or not their actions are aligned with those values.

- Have workers identify and prioritize individual values and have them discuss how their values can help them to want to work more safely.

YOUR NOTES PAGE

BRINGING IT TOGETHER

As I mentioned in the beginning of this book, several methods would be used to make your reading a more valuable learning experience. The Prologue helps to set the stage and introduce the *7C's* and their interactive nature. And together with the illustrations, the text helps to bring perspective to each principle.

Each of the *7C's* is fundamentally sound. And each one draws support from the other principles, even when you may not be aware that they're working together. In this way, the *7C's* could be further explored and explained as complementary principles, but time, space, and the nature of this book do not permit this discussion.

Scalable Learning at Your Fingertips

****

Have David Sarkus Deliver This Training and More Right to You and Your People at Any Time or Place!

To Test Drive Our Courses Please Visit

****

VirtualDavidSarkus.com

****

SIXTY SECONDS WITH SAFETY IN IT

Sixty seconds for you to choose,

Sixty seconds for you to lose,

Sixty seconds just one minute,

Just one minute with safety in the center of it.

One minute to do my best,

One minute to put safety to the test,

I only have one minute.

One minute to convince myself that working safely is to do what's best.

To help those around me know that working safely is their very best.

One minute that just might put them to the test.

One minute, that's sixty seconds, that's it.

Sixty seconds that could hold all of eternity with our safety wrapped up in it!

Copyright © 2001 - 2006, David J. Sarkus

DAVID AS A SPEAKER

"David Sarkus is an excellent communicator who is making a difference for people
all around the country!"

**— Willie Jolley
Certified Speaking Professional
Toastmasters International
Motivational Speaker of the Year
Selected as One of the Five Best Speakers
In the World**

"David Sarkus is an outstanding speaker! In my 20 plus years as a safety professional, I've seen just about all of the best speakers in this field and would rate David at the top. If you want to get your people excited and keep them moving in the right direction — they need to see and hear David Sarkus!"

**— Wally Cackowski, MS
Certified Safety Professional
Technical Trainer • Dominion Transmission**

"Just wanted to thank you for a phenomenal presentation. I walked away from your session with many ideas."

**— Jeremy A. Kullman • Safety Engineer
Tilcon Connecticut Inc.**

"The presentation you provided during National Safety Month at the Beaver Valley Power Station was OUTSTANDING!"

— John Kowalski
Certified Safety Professional
Industrial Safety Engineer
Duquesne Light Company

"David Sarkus is an outstanding person and speaker. We use him on a regular basis to keep our people excited and involved with safety. Through David, we know that safety is all about productivity improvement."

— Blair E. Merkel • Vice President
Sithe Energies

"David has the unique ability to relate effectively with people from all walks — academics to shop-floor workers. Combine that with his strong knowledge of both safety and psychology and you have an exceptional speaker and consultant."

— John Kamp, Ph.D. • Vice President
Reid Psychological Systems

ADDITIONAL SERVICES

David Sarkus International provides a full menu of value-added services ranging from keynotes, seminars, training camps, use of various assessment tools, and interventions such as behavior-based safety processes which make further use of the *7C's.*

In addition, audiotapes, workbooks, and audiovisual products are available for purchase. Bulk discounts are available as well as company discounts when you bring David in as a speaker or consultant. Call or write for more information.

SUCCESS STORIES

If you have a small or large success story related to the *7C's* please let us know in writing.

Championship Publishing
PO Box 137
Donora, PA 15033

www.DavidSarkus.com

800-240-4601

724-379-6439